ULTIMATE SUPERCARS

# CHEVROLET CORVETTE C8

By John Perritano

Kaleidoscope
Minneapolis, MN

## The Quest for Discovery Never Ends

..............................................

This edition first published in 2021 by Kaleidoscope Publishing, Inc.

No part of this publication may be reproduced in whole or in part without written permission of the publisher.

For information regarding permission, write to
Kaleidoscope Publishing, Inc.
6012 Blue Circle Drive
Minnetonka, MN 55343

Library of Congress Control Number
2020936095

ISBN
978-1-64519-260-2 (library bound)
978-1-64519-328-9 (ebook)

Text copyright © 2021 by Kaleidoscope Publishing, Inc. All-Star Sports, Bigfoot Books, and associated logos are trademarks and/or registered trademarks of Kaleidoscope Publishing, Inc.

Printed in the United States of America.

**FIND ME IF YOU CAN!**

Bigfoot lurks within one of the images in this book. It's up to you to find him!

# TABLE OF CONTENTS

**Chapter 1:** Tommy Milner Goes for a Spin ............... 4

**Chapter 2:** Car of the Year ............... 10

**Chapter 3:** Not Just About Speed ............... 16

**Chapter 4:** Top On, Top Off ............... 24

*Beyond the Book* ............... *28*

*Research Ninja* ............... *29*

*Further Resources* ............... *30*

*Glossary* ............... *31*

*Index* ............... *32*

*Photo Credits* ............... *32*

*About the Author* ............... *32*

## Chapter 1
# Tommy Milner Goes for a Spin

It's November 2019. Tommy Milner can't wait to get to Daytona for the big race. He first wants to show off his new ride in Atlanta. It's the Corvette C8.R.

Corvette fans from across the South pack the Atlanta track. Milner is about to put the C8.R through its paces. He has been racing Corvettes for

10 years. He won 24 Hours of Le Mans in 2011 and 2015. During both races he drove a Corvette C6.R.

The C8 is much different. Just look at it. The car's headlights look like alien eyes. Its rear **spoiler** is sleek and slanted.

The C8 looks like the future.

It also sounds like the future. The crowd can hear it when Milner revs the engine. Every **piston** moves as it should. Fans mill around the car. They ask questions. They take pictures.

# PARTS OF A
# CORVETTE C8

**FUN FACT**
The door handles are hidden when the door closes.

Hard top shown here; also comes in **convertible**

Air intakes

Michelin Pilot Sport tires

Milner puts on the seat belt. He's going to need it. He shifts. He makes a right turn. He steps on the gas. The C8 flies onto the track. Fans cheer. They snap pics.

You can hear the *oohs* and *aahs*.

The C8 handles beautifully as it makes tight turns. The engine hums with each gear.

Milner will race the C8.R for the first time in January. The race will be at the Rolex 24. The race is also called the **24 Hours of Daytona**.

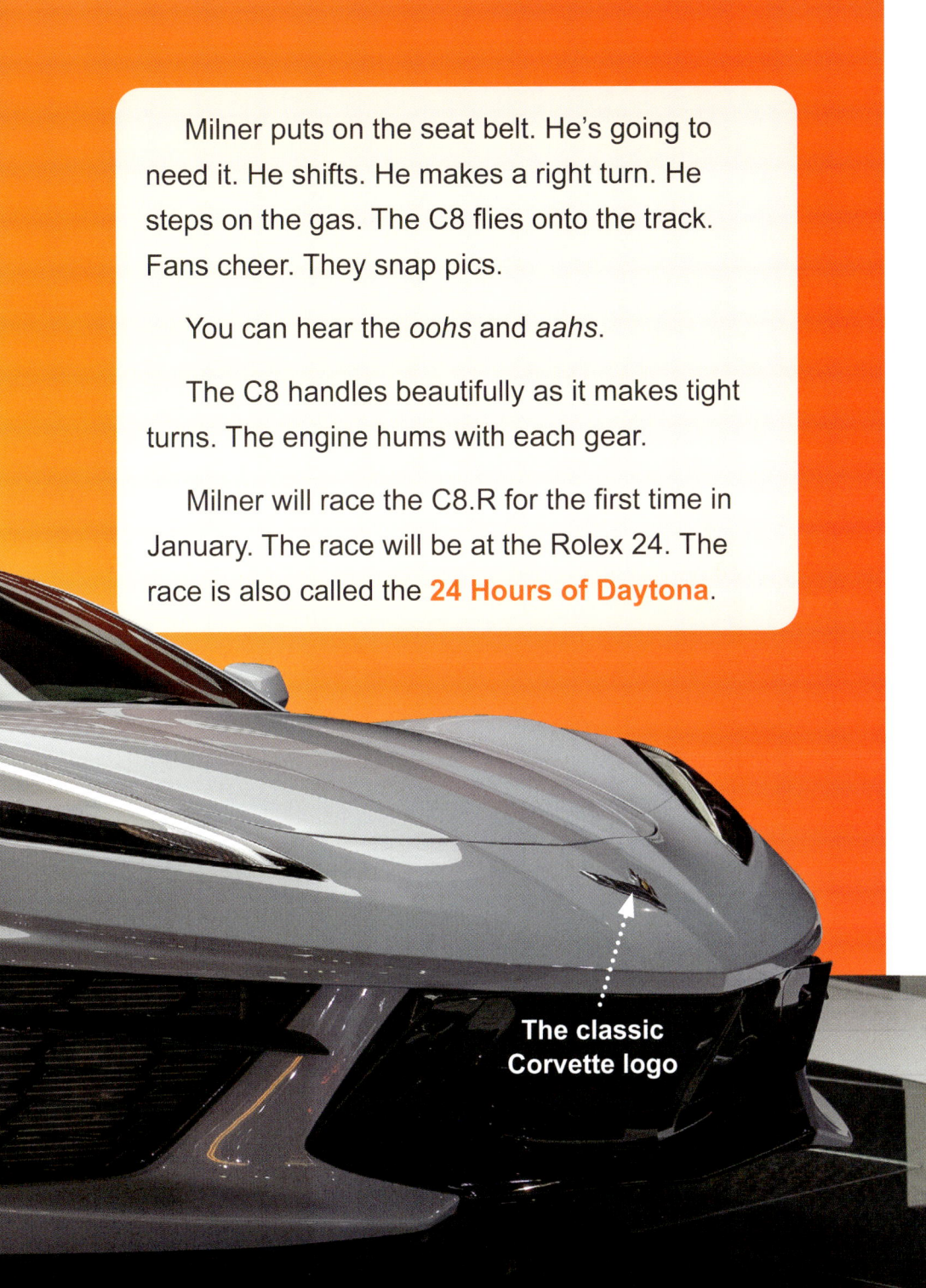

**The classic Corvette logo**

As Milner steers around the track it's hard not to hear the muscle of the C8's 500-**horsepower** engine. Fans are bug-eyed. They can't believe what they are seeing and hearing.

The C8 can sure fly. It handles well as Milner **downshifts** through the turns. The sun reflects off the car's silver and white paint.

The 24 Hours of Daytona will be the ultimate test. Will the C8 perform well? Is it reliable? Will Milner take home another trophy?

Push-button transmission

Here's a look at the convertible version.

## Chapter 2
## Car of the Year

Tom Peters liked to draw as a child. He drew people. He drew plants. He drew animals. He also drew cars. One day he got off the bus with his friends. He saw a silver flying-saucer coming in for a landing.

*A classic 1963 Corvette Stingray*

This was no flying saucer. It was a '63 Stingray Corvette. Peters fell in love in that moment with the car and with the Vette. In 1980, he went to work for **General Motors** as a designer. One of his first jobs was to work on a mid-engine Corvette. He had to place its engine between the front and rear axles.

**The Corvette logo includes a checkered flag from racing. The fleur-de-lis (far right) is for Chevrolet's French heritage.**

Peters went to work, remembering the day when the silver Stingray came into view. GM would later promote Peters to its chief of design. He helped shape the C6 Corvette and the C7.

But it would be the C8 Stingray that would cap his career. The Batmobile-like car would be the last he would design. People couldn't wait to get a look at it. Some called it "mid-engine" supercar.

Corvettes became larger and bolder by 2015, when this C6 was new.

# WHERE THE CORVETTE IS MADE

**Detroit, Michigan:** Chevrolet Headquarters

**Bowling Green, Kentucky:** Where the Corvette C8 is assembled

**FUN FACT**
Corvettes are often seen on the twisting track at Road America in Wisconsin.

14

What people saw they loved. *Motor Trend* magazine named the C8 2020's Car of the Year. The magazine called the eighth-generation Vette a "game changer."

It handled easily even at high speeds and around corners. The engine placement made the car perform better than any other Vette. The C8 beat out 10 heavy-hitting performance cars on a figure-eight track.

"We've been waiting so long for this car. I felt like a kid on Christmas morning," wrote one writer.

## WHERE'S THE ENGINE?

One of the reasons the C8 can go very fast is because its engine is in the middle of the car. The engine puts weight on the rear tires. The increased weight makes the tires grip the road better. The car can put its power to the ground. The C7 only has 1,750 pounds on its rear tires. That's because its engine is in the front. The C8 has 2,210 pounds on its rear tires.

## Chapter 3
## Not Just About Speed

Stacy got into her new C8. She couldn't believe her good luck. Her mom loves Vettes. Her dad too. They drove it home yesterday.

"This is awesome," Stacy says as her mom starts the car.

"Buckle up," her mom says, as she shifts into first gear. "We're going for a ride."

## WHAT'S INSIDE?

The interior of a Corvette was never its biggest draw. That has all changed with the C8. Its dashboard is laid out in 3D. The cockpit wraps around the driver.

The squared-off steering wheel is wrapped in either leather or Alcantara. Alcantara is a suede-like microfiber. The car's seats come in three different designs. The GT1 is sporty and comfortable. The GT2 is great for long road trips.

*Corvettes love to hit the highway!*

It's a great time to be a Corvette lover. The C8 is fast. It can go from 0 to 60 miles per hour (96.56 kph) in 2.8 seconds. It's a Vette after all.

Stacy's mom gets on the freeway. Stacy could feel the power of the C8's **V8**.

The seats are based on those used by race car drivers.

**FUN FACT**
Corvette buyers can choose from among 13 interior colors in 2020.

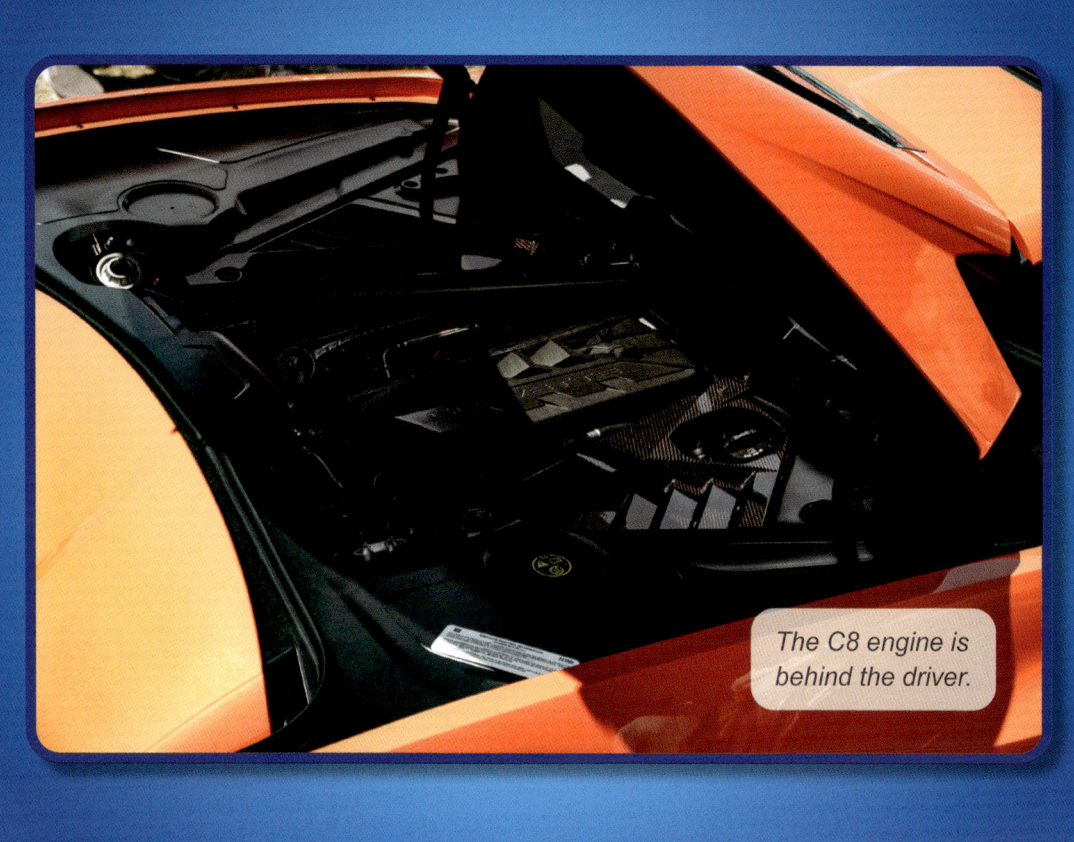

The C8 engine is behind the driver.

But Corvettes have never just been about speed. They are quality cars.

The C8's futuristic interior is top-end. Stacy feels comfortable inside. The leather seats slope back. She's not that tall but can see over the dashboard. That's because the engine is not in the front. It's behind the seats.

# THE CORVETTE C8
## IN DETAIL

Height: 4.05 feet, (1.23 m)

Length: 15 feet, 2 inches (4.62 m)

Width: 6 feet, 4 inches (1.93 m)

**LENGTH:** 15 feet, 2 inches (4.62 m)

**WEIGHT:** 3,336 pounds (1,527 kg)

**TOP SPEED:** 194 miles per hour (312.21 kph)

**TIME FROM 0-60 MPH:** 2.8 seconds

**COST: $74,295**

Stacy's mom grips the steering wheel. It's part oval and part square. It's nothing like a traditional steering wheel.

"This bad boy rides quiet," Stacy says.

"Doesn't it?" Stacy's mom agrees. "You can hear very little road noise."

**FUN FACT**
Buyers can order the steering wheel made of carbon fiber!

The C8 is fast. Very fast! Its top speed is 194 miles an hour. Stacy's mom hits the gas. She speeds effortlessly around a curve.

"Geez, Mom," Stacy says. "This baby can sure hug the corners."

As well it should. The C8's rear-deck spoiler creates a lot of **downforce**. It allows a car to travel faster around a corner.

## THE UPSIDE OF DOWNFORCE

Downforce pushes down on a car allowing the tires to grip the road better. The image below shows how air (shown as white smoke) passes smoothly over a car that is designed to have downforce.

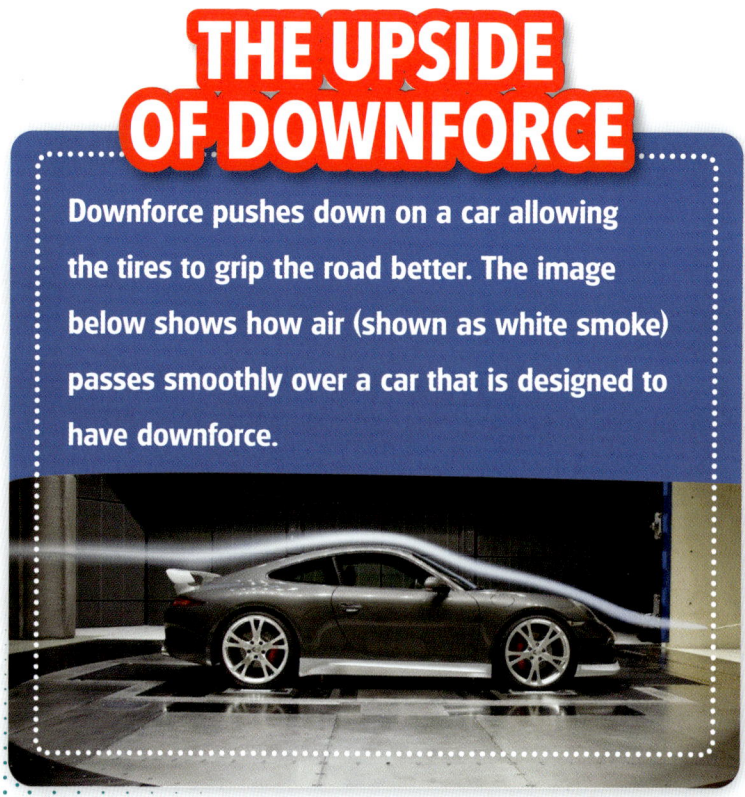

*The rear spoiler helps the Corvette cut through the air.*

## Chapter 4
# Top On, Top Off

"You and Dad should have gotten the convertible," Stacy says.

"We thought about it. But Dad wanted the coupe."

Corvette has always been a convertible-first car. The C8 continues that tradition. It has a **retractable** roof. First, the back slowly lifts up. Then, the roof pulls back. It flows seamlessly into the body.

Convertible or hardtop: Which would you choose?

The roof is powered by six motors. It can be taken off at speeds up to 30 miles per hour (48.28 kph). The convertible weighs 77 pounds (34.9 kg) more than the **hardtop**. The extra weight doesn't affect the car's performance.

## WHY A MID-ENGINE?

The C8's mid-engine makes it easy for the driver to see out the front window. It also directs more power to the rear wheels. It allowed the car's designers to put the windshield over the front wheels. That helps create more downforce in the front of the car.

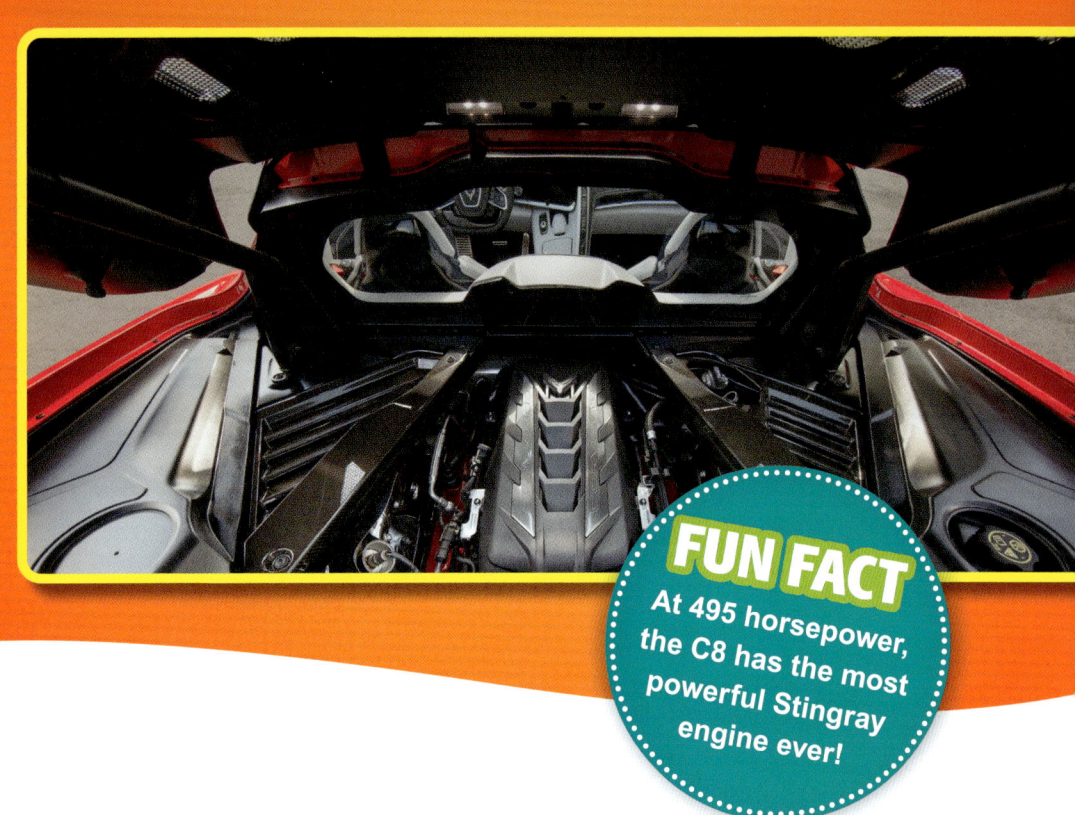

**FUN FACT**
At 495 horsepower, the C8 has the most powerful Stingray engine ever!

"I can't wait to take this to the grocery store," Stacy's mom says.

"Where would you put the bags?" Stacy asks.

"There's plenty of room under the hood," her mom says. "That's because there is no engine in front."

The C8 has a lot of cargo space for a Vette. There's a deep well in the front of the vehicle. Just pop the hood open. There's also a trunk in the rear. You can put two bags of golf clubs inside.

The C8's storage system makes it a great car to go on vacation with.

The C8 Stingray is Corvette at its best. Its control panel is easy to use. It looks as if it should be in a fighter jet. Each button is exactly where it needs to be.

"I love this car," Stacy says. "Will you drive me to school tomorrow?"

Her mom laughs. "It's not nice to show off."

Corvette buyers can choose a paint color and a wheel style, too.

# BEYOND THE BOOK

After reading the book, it's time to think about what you learned. Try the following exercises to jumpstart your ideas.

## RESEARCH

**FIND OUT MORE.** Where would you go to find out more about your favorite cars? Find out what company makes the car and locate its website. What information do the companies provide? What other sources of car information can you find?

## CREATE

**GET ARTISTIC.** Cars start with creative artists and designers. Time for you to take a shot! Get art materials and create a great, new car. Will you make it a sports car? A sedan? A race car? What colors will you paint it? What features will you give it? Let your imagination go for a spin!

## DISCOVER

**DIG DEEPER.** Corvettes have often been used in movies and TV. Do some research and find out some of the examples. If it's okay with your folks, watch some of the shows. How did the filmmakers use the cars? Were they a big part of the plots? What were your favorites?

## GROW

**GO TO A CAR SHOW.** Car shows are a great way to see lots of cool cars up-close. Check your local events calendar, or ask at a car dealer for upcoming events. You can find shows of old cars and new cars, sports cars and classic cars. Go to a show and find a new favorite car to love!

# RESEARCH NINJA

Visit *www.ninjaresearcher.com/2602* to learn how to take your research skills and book report writing to the next level!

## RESEARCH

**DIGITAL LITERACY TOOLS**

### SEARCH LIKE A PRO
Learn about how to use search engines to find useful websites.

### FACT OR FAKE?
Discover how you can tell a trusted website from an untrustworthy resource.

### TEXT DETECTIVE
Explore how to zero in on the information you need most.

### SHOW YOUR WORK
Research responsibly—learn how to cite sources.

## WRITE

### GET TO THE POINT
Learn how to express your main ideas.

### PLAN OF ATTACK
Learn prewriting exercises and create an outline.

**DOWNLOADABLE REPORT FORMS**

# Further Resources

## BOOKS

Cruz, Calvin. *Chevrolet Corvette Z06*. Minneapolis, MN: Bellwether Media, 2016.

Kingston, Seth. *The History of Corvettes*. New York, NY: PowerKids Press, 2019.

Peppas, Lynn. *Corvette (Supercars)*. Toronto, Ont.: Crabtree Publishing, 2010.

## WEBSITES

### FACTSURFER

Factsurfer.com gives you a safe, fun way to find more information.

1. Go to www.factsurfer.com.
2. Enter "Chevrolet Corvette C8" into the search box and click 🔍
3. Select your book cover to see a list of related websites.

**24 Hours of Daytona:** a 24-hour sports car endurance race held annually at Daytona International Speedway in Daytona Beach, Florida.

**convertible:** a car with a roof that folds down.

**downforce:** force produced by air resistance and gravity that presses down on a moving vehicle, giving the vehicle more stability.

**downshift:** change to a lower gear in a motor vehicle.

**General Motors:** one of the three big US carmakers; General Motors owns Buick, Cadillac, GMC, and Chevrolet.

**hardtop:** the hard roof on top of a car.

**horsepower:** one horsepower is the power it takes to lift 550 pounds one foot in one second.

**piston:** part of a car's engine that transfers force from expanding gas in a cylinder to the crankshaft that allows the car to move.

**retractable:** having the ability to withdraw inside.

**spoiler:** an aerodynamic flap on the back of a car that is designed to "spoil" the unfavorable air movement across the body of a vehicle.

**V8:** A V8 engine has eight cylinders in the shape of a V.

# Index

24 Hours of Daytona, 7, 8
24 Hours of Le Mans, 5
Batmobile, 12
cargo space, 26
convertible, 6, 9, 24, 25
Corvette Stingray, 10, 11, 12
Daytona, 4
downforce, 22, 25
interior, 17, 18, 19

Michigan, 13
mid-engine, 11, 12, 25
Milner, Tommy, 4, 5, 7, 8
*Motor Trend* magazine, 15
Peters, Tom, 10, 11, 12
steering wheel, 17, 21
Road America, 14
Rolex 24, 7
Wisconsin, 14

## PHOTO CREDITS

The images in this book are reproduced through the courtesy of: Chevrolet: 4, 9, 14, 16, 21, 26, 27. Shutterstock: Betto Rodrigues 6; Terry Butler Photography 8, 18, 20; Natursports 12; Christopher Lyzcen 19, 23, 25. Wikimedia: sv1ambo 10.
**Cover:** Lawrence Carmichael/Shutterstock (car); Dmitry Rukhlenko/Shutterstock (background, top); zhao jiankang/Shutterstock (background, bottom).

# About the Author

John Perritano is an award-winning journalist, author, and editor from Southbury, Connecticut. He has authored numerous books and articles on subjects such as science, technology, history, and current events. He holds a master's degree in American History from Western Connecticut State University.